iSearch

Allyn & Bacon

Boston   New York   San Francisco
Mexico City   Montreal   Toronto   London   Madrid   Munich   Paris
Hong Kong   Singapore   Tokyo   Cape Town   Sydney

Copyright © 2003 Pearson Education, Inc.

All rights reserved. No part of the material protected by this copyright notice may be reproduced or utilized in any form or by any means, electronic or mechanical, including photocopying, recording, or by any information storage and retrieval system, without the written permission of the copyright owner.

To obtain permission(s) to use material from this work, please submit a written request to Allyn and Bacon, Permissions Department, 75 Arlington Street, Suite 300, Boston, MA 02116 or fax your request to 617-848-7320.

ISBN   0-205-38148-0

Printed in the United States of America

10 9 8 7 6 5 4 3 2   08 07 06 05 04 03 02

Contents

Introduction iv

PART 1 CONDUCTING ONLINE RESEARCH 1

    Finding Sources: Search Engines
       and Subject Directories 1

    Evaluating Sources on the Web 9

    Documentation Guidelines for Online Sources 16

PART 2 CONTENTSELECT 22

    What Is ContentSelect? 22

    How to Use ContentSelect 23

    ContentSelect Search Tips 24

GLOSSARY 27

Introduction

Your professor assigns a ten-page research paper that's due in two weeks—and you need to make sure you have up-to-date, credible information. Where do you begin? Today, the easiest answer is the Internet—because it can be so convenient and there is so much information out there. But therein lies part of the problem. How do you know if the information is reliable and from a trustworthy source?

iSearch is designed to help you select and evaluate research from the Web to help you find the best and most credible information you can. Throughout this guide, you'll find:

- **A practical and to-the-point discussion of search engines.** Find out which search engines are likely to get you the information you want and how to phrase your searches for the most effective results.
- **Detailed information on evaluating online sources.** Locate credible information on the Web and get tips for thinking critically about Web sites.
- **Citation guidelines for Web resources.** Learn the proper citation guidelines for Web sites, email messages, listservs, and more.
- **A quick guide to ContentSelect.** All you need to know to get started with ContentSelect, a research database that gives you immediate access to hundreds of scholarly journals and other popular publications, such as *Newsweek*.

So before running straight to your browser, take the time to read through this copy of *iSearch* and use it as a reference for all of your Web research needs.

PART 1

Conducting Online Research

Finding Sources: Search Engines and Subject Directories

Your professor has just given you an assignment to give a five minute speech on the topic "gun control." After a (hopefully brief) panic attack, you begin to think of what type of information you need before you can write the speech. To provide an interesting introduction, you decide to involve your class by taking a straw poll of their views for and against gun control, and to follow this up by giving some statistics on how many Americans favor (and oppose) gun control legislation and then by outlining the arguments on both sides of the issue. If you already know the correct URL for an authoritative Web site like Gallup Opinion Polls (www.gallup.com) or other sites you are in great shape! However, what do you do when you don't have a clue as to which Web site would have information on your topic? In these cases, many, many people routinely (and mistakenly) go to Yahoo! and type in a single term (e.g., guns). This approach is sure to bring first a smile to your face when the results offer you 200,874 hits on your topic, but just as quickly make you grind your teeth in frustration when you start scrolling down the hit list and find sites that range from gun dealerships, to reviews of the video "Young Guns," to aging fan sites for "Guns and Roses."

Finding information on a specific topic on the Web is a challenge. The more intricate your research need, the more difficult it is to find the one or two Web sites among the billions that feature the information you want. This section is designed to help you to avoid frustration and to focus in on the right site for your research by using search engines, subject directories, and meta-sites.

Search Engines

Search engines (sometimes called search services) are becoming more numerous on the Web. Originally, they were designed to help users search the Web by topic. More recently, search engines have added features which enhance their usefulness, such as searching a particular part of the Web (e.g., only sites of educational institutions—dot.edu), retrieving just one site which the search engine touts as most relevant (like Ask Jeeves {www.aj.com}), or retrieving up to 10 sites which the search engine rank as most relevant (like Google {www.google.com}).

Search Engine Defined

According to Cohen (1999):

> "A search engine service provides a searchable database of Internet files collected by a computer program called a wanderer, crawler, robot, worm, or spider. Indexing is created from the collected files, and the results are presented in a schematic order. There are no selection criteria for the collection of files.
>
> A search service therefore consists of three components: (1) a spider, a program that traverses the Web from link to link, identifying and reading pages; (2) an index, a database containing a copy of each Web page gathered by the spider; and (3) a search engine mechanism, software that enables users to query the index and then returns results in a schematic order (p. 31)."

One problem students often have in their use of search engines is that they are deceptively easy to use. Like our example "guns," no matter what is typed into the handy box at the top, links to numerous Web sites appear instantaneously, lulling students into a false sense of security. Since so much was retrieved, surely SOME of it must be useful. WRONG! Many Web sites retrieved will be very light on substantive content, which is not what you need for most academic endeavors. Finding just the right Web site has been likened to finding diamonds in the desert.

As you can see by the definition above, one reason for this is that most search engines use indexes developed by machines. Therefore they are indexing terms not concepts. The search engine cannot tell the difference between the keyword "crack" to mean a split in the sidewalk and "crack" referring to crack cocaine. To use search engines properly takes some skill, and this chapter will provide tips to help you use search engines more effectively. First, however, let's look at the different types of search engines with examples:

TYPES OF SEARCH ENGINES

TYPE	DESCRIPTION	EXAMPLES
1st Generation	• Non-evaluative, do not evaluate results in terms of content or authority. • Return results ranked by relevancy alone (number of times the term(s) entered appear, usually on the first paragraph or page of the site)	AltaVista (www.altavista.com/) Excite (www.excite.com) HotBot (www.HotBot.com) Infoseek (guide.infoseek.com) Ixquick Metasearch (ixquick.com) Lycos (www.lycos.com)
2nd Generation	• More creative in displaying results. • Results are ordered by characteristics such as: concept, document type, Web site, popularity, etc. rather than relevancy.	Ask Jeeves (www.aj.com/) Direct Hit (www.directhit.com/) Google! (www.google.com/) HotLinks (www.hotlinks.com/) Simplifind (www.simpli.com/) SurfWax (www.surfwax.com/) Also see Meta-Search engines below. EVALUATIVE SEARCH ENGINES About.Com (www.about.com) WebCrawler (www.webcrawler.com)
Commercial Portals	• Provide additional features such as: customized news, stock quotations, weather reports, shopping, etc. • They want to be used as a "one stop" Web guide. • They profit from prominent advertisements and fees charged to featured sites.	GONetwork (www.go.com/) Google Web Directory (directory.google.com/) LookSmart (www.looksmart.com/) My Starting Point (www.stpt.com/) Open Directory Project (dmoz.org/) NetNow (www.inetnow.com) Yahoo! (www.yahoo.com/)
Meta-Search Engines	Run searches on multiple search engines.	There are different types of meta-search engines. See the next 2 boxes.

(continued)

TYPES OF SEARCH ENGINES, *continued*

TYPE	DESCRIPTION	EXAMPLES
Meta-Search Engines *Integrated Result*	• Display results for search engines in one list. • Duplicates are removed. • Only portions of results from each engine are returned.	Beaucoup.com (www.beaucoup.com/) Highway 61 (www.highway61.com) Cyber411(www.cyber411.com/) Mamma (www.mamma.com/) MetaCrawler (www.metacrawler.com/) Visisimo (www.vivisimo.com) Northern Light (www.nlsearch.com/) SurfWax (www.surfwax.com)
Meta-Search Engines *Non-Integrated Results*	• Comprehensive search. • Displays results from each search engine in separate results sets. • Duplicates remain. • You must sift through all the sites.	Dogpile (www.dogpile.com) Global Federated Search (jin.dis.vt.edu/fedsearch/) GoHip (www.gohip.com) Searchalot (www.searchalot.com) 1Blink (www.1blink.com) ProFusion (www.profusion.com/)

QUICK TIPS FOR MORE EFFECTIVE USE OF SEARCH ENGINES

1. Use a search engine:
   - When you have a narrow idea to search.
   - When you want to search the full text of countless Web pages
   - When you want to retrieve a large number of sites
   - When the features of the search engine (like searching particular parts of the Web) help with your search

2. Always use Boolean Operators to combine terms. Searching on a single term is a sure way to retrieve a very large number of Web pages, few, if any, of which are on target.
   - Always check search engine's HELP feature to see what symbols are used for the operators as these vary (e.g., some engines use the & or + symbol for AND).
   - Boolean Operators include:
     AND to narrow search and to make sure that **both** terms are included
       e.g:, children AND violence
     OR to broaden search and to make sure that **either** term is included
       e.g., child OR children OR juveniles
     NOT to **exclude** one term
       e.g., eclipse NOT lunar

3. Use appropriate symbols to indicate important terms and to indicate phrases (Best Bet for Constructing a Search According to Cohen (1999): Use a plus sign (+) in front of terms you want to retrieve: +solar +eclipse. Place a phrase in double quotation marks: "solar eclipse" Put together: "+solar eclipse" "+South America").

4. Use word stemming (a.k.a. truncation) to find all variations of a word (check search engine HELP for symbols).
    - If you want to retrieve child, child's, or children use child* (some engines use other symbols such as !, #, or $)
    - Some engines automatically search singular and plural terms, check HELP to see if yours does.

5. Since search engines only search a portion of the Web, use several search engines or a meta-search engine to extend your reach.

6. Remember search engines are generally mindless drones that do not evaluate. Do not rely on them to find the best Web sites on your topic, use *subject directories* or meta-sites to enhance value (see below).

Finding Those Diamonds in the Desert: Using Subject Directories and Meta-sites

Although some search engines, like WebCrawler (www.webcrawler.com) do evaluate the Web sites they index, most search engines do not make any judgment on the worth of the content. They just return a long—sometimes very long—list of sites that contained your keyword. However, *subject directories* exist that are developed by human indexers, usually librarians or subject experts, and are defined by Cohen (1999) as follows:

> "A subject directory is a service that offers a collection of links to Internet resources submitted by site creators or evaluators and organized into subject categories. Directory services use selection criteria for choosing links to include, though the selectivity varies among services (p. 27)."

World Wide Web Subject directories are useful when you want to see sites on your topic that have been reviewed, evaluated, and selected for their authority, accuracy, and value. They can be real time savers for students, since subject directories weed out the commercial, lightweight, or biased Web sites.

Metasites are similar to subject directories, but are more specific in nature, usually dealing with one scholarly field or discipline. Some examples of subject directories and meta-sites are found in the table on the next page.

SMART SEARCHING—SUBJECT DIRECTORIES AND META-SITES

TYPES—SUBJECT DIRECTORIES	EXAMPLES
General, covers many topics	Access to Internet and Subject Resources (www2.lib.udel.edu/subj/) Best Information on the Net (BIOTN) (http://library.sau.edu/bestinfo/) Federal Web Locator (www.infoctr.edu/fwl/) Galaxy (galaxy.einet.net) INFOMINE: Scholarly Internet Resource Collections (infomine.ucr.edu/) InfoSurf: Resources by Subject (www.library.ucsb.edu/subj/) Librarian's Index to the Internet (www.lii.org/) Martindale's "The Reference Desk" (www-sci.lib.uci.edu/HSG/ref.html) PINAKES: A Subject Launchpad (www.hw.ac.uk/libWWW/irn/pinakes/pinakes.html) Refdesk.com (www.refdesk.com) Search Engines and Subject Directories (College of New Jersey) (www.tcnj.edu/~library/research/internet_search.html) Scout Report Archives (www.scout.cs.wisc.edu/archives) Selected Reference Sites (www.mnsfld.edu/depts/lib/mu~ref.html) WWW Virtual Library (http://vlib.org)
Subject Oriented	
• Communication Studies	The Media and Communication Studies Site (www.aber.ac.uk/media) University of Iowa Department of Communication Studies (www.uiowa.edu/~commstud/resources)
• Cultural Studies	Sara Zupko's Cultural Studies Center (www.popcultures.com)
• Education	Educational Virtual Library (www.csu.edu.au/education/library.html) ERIC [Education ResourcesInformation Center] (ericir.sunsite.syr.edu/) Kathy Schrock's Guide for Educators (kathyschrock.net/abceval/index.htm)
• Journalism	Journalism Resources (bailiwick.lib.uiowa.edu/journalism/) Journalism and Media Criticism page (www.chss.montclair.edu/english/furr/media.html)
• Literature	Norton Web Source to American Literature (www.wwnorton.com/naal) Project Gutenberg [Over 3,000 full text titles] (www.gutenberg.net)

SMART SEARCHING, *continued*

TYPES—SUBJECT DIRECTORIES	EXAMPLES
• Medicine & Health	PubMed [National Library of Medicine's index to Medical journals, 1966 to present] (www.ncbi.nlm.nih.gov/PubMed/) RxList: The Internet Drug Index (rxlist.com) Go Ask Alice (www.goaskalice.columbia.edu) [Health and sexuality]
• Technology	CNET.com (www.cnet.com)

Choose subject directories to ensure that you are searching the highest quality Web pages. As an added bonus, subject directories periodically check Web links to make sure that there are fewer dead ends and out-dated links.

Another closely related group of sites are the *Virtual Library sites*, also referred to as Digital Library sites (see the table below). Hopefully, your campus library has an outstanding Web site for both on-campus and off-campus access to resources. If not, there are several

VIRTUAL LIBRARY SITES

PUBLIC LIBRARIES	
• Internet Public Library	www.ipl.org
• Library of Congress	lcweb.loc.gov/homepage/lchp.html
• New York Public Library	www.nypl.org
University/College Libraries	
• Bucknell	jade.bucknell.edu/
• Case Western	www.cwru.edu/uclibraries.html
• Dartmouth	www.dartmouth.edu/~library
• Duke	www.lib.duke.edu/
• Franklin & Marshall	www.library.fandm.edu
• Harvard	www.harvard.edu/museums/
• Penn State	www.libraries.psu.edu
• Princeton	infoshare1.princeton.edu
• Stanford	www.slac.stanford.edu/FIND/spires.html
• ULCA	www.library.ucla.edu
Other	
• Perseus Project [subject specific—classics, supported by grants from corporations and educational institutions]	www.perseus.tufts.edu

virtual library sites that you can use, although you should realize that some of the resources would be subscription based, and not accessible unless you are a student of that particular university or college. These are useful because, like the subject directories and meta-sites, experts have organized Web sites by topic and selected only those of highest quality.

You now know how to search for information and use search engines more effectively. In the next section, you will learn more tips for evaluating the information that you found.

BIBLIOGRAPHY FOR FURTHER READING

Books

Basch, Reva. (1996). Secrets of the Super Net Searchers.

Berkman, Robert I. (2000). *Find It Fast: How to Uncover Expert Information on Any Subject Online or in Print.* NY: HarperResource.

Glossbrenner, Alfred & Glossbrenner, Emily. (1999). *Search Engines for the World Wide Web,* 2nd Ed. Berkeley, CA: Peachpit Press.

Hock, Randolph, & Berinstein, Paula.. (1999). *The Extreme Searcher's Guide to Web Search Engines: A Handbook for the Serious Searcher.* Information Today, Inc.

Miller, Michael. *Complete Idiot's Guide to Yahoo!* (2000). Indianapolis, IN: Que.

Miller, Michael. *Complete Idiot's Guide to Online Search Secrets.* (2000). Indianapolis, IN: Que.

Paul, Nora, Williams, Margot, & Hane, Paula. (1999). *Great Scouts!: Cyber-Guides for Subject Searching on the Web.* Information Today, Inc.

Radford, Marie, Barnes, Susan, & Barr, Linda (2001). *Web Research: Selecting, Evaluating, and Citing* Boston. Allyn and Bacon.

Journal Articles

Cohen, Laura B. (1999, August). The Web as a research tool: Teaching strategies for instructors. *CHOICE Supplement 3,* 20–44.

Cohen, Laura B. (August 2000). Searching the Web: The Human Element Emerges. *CHOICE Supplement 37,* 17–31.

Introna, Lucas D., & Nissenbaum, Helen. (2000). Shaping the web: Why the politics of search engines matters. The Information Society, Vol. 16, No. 3, pp. 169–185.

Evaluating Sources on the Web

Congratulations! You've found a great Web site. Now what? The Web site you found seems like the perfect Web site for your research. But, are you sure? Why is it perfect? What criteria are you using to determine whether this Web site suits your purpose?

Think about it. Where else on earth can anyone "publish" information regardless of the *accuracy, currency,* or *reliability* of the information? The Internet has opened up a world of opportunity for posting and distributing information and ideas to virtually everyone, even those who might post misinformation for fun, or those with ulterior motives for promoting their point of view. Armed with the information provided in this guide, you can dig through the vast amount of useless information and misinformation on the World Wide Web to uncover the valuable information. Because practically anyone can post and distribute their ideas on the Web, you need to develop a new set of *critical thinking skills* that focus on the evaluation of the quality of information, rather than be influenced and manipulated by slick graphics and flashy moving java script.

Before the existence of online sources, the validity and accuracy of a source was more easily determined. For example, in order for a book to get to the publishing stage, it must go through many critiques, validation of facts, reviews, editorial changes and the like. Ownership of the information in the book is clear because the author's name is attached to it. The publisher's reputation is on the line too. If the book turns out to have incorrect information, reputations and money can be lost. In addition, books available in a university library are further reviewed by professional librarians and selected for library purchase because of their accuracy and value to students. Journal articles downloaded or printed from online subscription services, such as Infotrac, ProQuest, EbscoHost, or other fulltext databases, are put through the same scrutiny as the paper versions of the journals.

On the World Wide Web, however, Internet service providers (ISPs) simply give Web site authors a place to store information. The Web site author can post information that may not be validated or tested for accuracy. One mistake students typically make is to assume that all information on the Web is of equal value. Also, in the rush to get assignments in on time, students may not take the extra time to make sure that the information they are citing is accurate. It is easy just to cut and paste without really thinking about the content in a critical way. However, to make sure you are gathering accurate information and to get the best grade on your assignments, it is vital that you develop your critical ability to sift through the dirt to find the diamonds.

Evaluating Web Sites Using
Five Criteria to Judge Web Site Content

Accuracy—How reliable is the information?

Authority—Who is the author and what are his or her credentials?

Objectivity—Does the Web site present a balanced or biased point of view?

Coverage—Is the information comprehensive enough for your needs?

Currency—Is the Web site up to date?

Use additional criteria to judge Web site content, including

- **Publisher, documentation, relevance, scope, audience, appropriateness of format,** and **navigation**
- Judging whether the site is made up of **primary (original) or secondary (interpretive) sources**
- Determining whether the information is **relevant** to your research

Web Evaluation Criteria

So, here you are, at this potentially great site. Let's go though some ways you can determine if this site is one you can cite with confidence in your research. Keep in mind, ease of use of a Web site is an issue, but more important is learning how to determine the validity of data, facts, and statements for your use. The five traditional ways to verify a paper source can also be applied to your Web source: *accuracy, authority, objectivity, coverage,* and *currency.*

Content Evaluation

Accuracy. Internet searches are not the same as searches of library databases because much of the information on the Web has not been edited, whereas information in databases has. It is your responsibility to make sure that the information you use in a school project is accurate. When you examine the content on a Web site or Web page, you can ask yourself a number of questions to determine whether the information is accurate.

1. Is the information reliable?
2. Do the facts from your other research contradict the facts you find on this Web page?
3. Do any misspellings and/or grammar mistakes indicate a hastily put together Web site that has not been checked for accuracy?
4. Is the content on the page verifiable through some other source? Can you find similar facts elsewhere (journals, books, or other online sources) to support the facts you see on this Web page?

5. Do you find links to other Web sites on a similar topic? If so, check those links to ascertain whether they back up the information you see on the Web page you are interested in using.
6. Is a bibliography of additional sources for research provided? Lack of a bibliography doesn't mean the page isn't accurate, but having one allows you further investigation points to check the information.
7. Does the site of a research document or study explain how the data was collected and the type of research method used to interpret the data?

If you've found a site with information that seems too good to be true, it may be. You need to verify information that you read on the Web by crosschecking against other sources.

Authority. An important question to ask when you are evaluating a Web site is, "Who is the author of the information?" Do you know whether the author is a recognized authority in his or her field? Biographical information, references to publications, degrees, qualifications, and organizational affiliations can help to indicate an author's authority. For example, if you are researching the topic of laser surgery citing a medical doctor would be better than citing a college student who has had laser surgery.

The organization sponsoring the site can also provide clues about whether the information is fact or opinion. Examine how the information was gathered and the research method used to prepare the study or report. Other questions to ask include:

1. Who is responsible for the content of the page? Although a webmaster's name is often listed, this person is not necessarily responsible for the content.
2. Is the author recognized in the subject area? Does this person cite any other publications he or she has authored?
3. Does the author list his or her background or credentials (e.g., Ph.D. degree, title such as professor, or other honorary or social distinction)?
4. Is there a way to contact the author? Does the author provide a phone number or email address?
5. If the page is mounted by an organization, is it a known, reputable one?
6. How long has the organization been in existence?
7. Does the URL for the Web page end in the extension .edu or .org? Such extensions indicate authority compared to dotcoms (.com), which are commercial enterprises. (For example, www.cancer.com takes you to an online drugstore that has a cancer information page; www.cancer.org is the American Cancer Society Web site.)

A good idea is to ask yourself whether the author or organization presenting the information on the Web is an authority on the subject. If the answer is no, this may not be a good source of information.

Objectivity. Every author has a point of view, and some views are more controversial than others. Journalists try to be objective by providing both sides of a story. Academics attempt to persuade readers by presenting a logical argument, which cites other scholars' work. You need to look for two sided arguments in news and information sites. For academic papers, you need to determine how the paper fits within its discipline and whether the author is using controversial methods for reporting a conclusion.

Authoritative authors situate their work within a larger discipline. This background helps readers evaluate the author's knowledge on a particular subject. You should ascertain whether the author's approach is controversial and whether he or she acknowledges this. More important, is the information being presented as fact or opinion? Authors who argue for their position provide readers with other sources that support their arguments. If no sources are cited, the material may be an opinion piece rather than an objective presentation of information. The following questions can help you determine objectivity:

1. Is the purpose of the site clearly stated, either by the author or the organization authoring the site?
2. Does the site give a balanced viewpoint or present only one side?
3. Is the information directed toward a specific group of viewers?
4. Does the site contain advertising?
5. Does the copyright belong to a person or an organization?
6. Do you see anything to indicate who is funding the site?

Everyone has a point of view. This is important to remember when you are using Web resources. A question to keep asking yourself is, What is the bias or point of *view* being expressed here?

Coverage. Coverage deals with the breadth and depth of information presented on a Web site. Stated another way, it is about how much information is presented and how detailed the information is. Looking at the site map or index can give you an idea about how much information is contained on a site. This isn't necessarily bad. Coverage is a criteria that is tied closely to *your* research requirement. For one assignment, a given Web site may be too general for your needs. For another assignment, that same site might be perfect. Some sites contain very little actual information because pages are filled with links to other sites. Coverage also relates to objectivity You should ask the following questions about coverage:

1. Does the author present both sides of the story or is a piece of the story missing?
2. Is the information comprehensive enough for your needs?
3. Does the site cover too much, too generally?
4. Do you need more specific information than the site can provide?
5. Does the site have an objective approach?

In addition to examining what is covered on a Web site, equally revealing is what is not covered. Missing information can reveal a bias in the material. Keep in mind that you are evaluating the information on a Web site for your research requirements.

Currency. Currency questions deal with the timeliness of information. However, currency is more important for some topics than for others. For example, currency is essential when you are looking for technology related topics and current events. In contrast, currency may not be relevant when you are doing research on Plato or Ancient Greece. In terms of Web sites, currency also pertains to whether the site is being kept up to date and links are being maintained. Sites on the Web are sometimes abandoned by their owners. When people move or change jobs, they may neglect to remove theft site from the company or university server. To test currency ask the following questions:

1. Does the site indicate when the content was created?
2. Does the site contain a last revised date? How old is the date? (In the early part of 2001, a university updated their Web site with a "last updated" date of 1901! This obviously was a Y2K problem, but it does point out the need to be observant of such things!)
3. Does the author state how often he or she revises the information? Some sites are on a monthly update cycle (e.g., a government statistics page).
4. Can you tell specifically what content was revised?
5. Is the information still useful for your topic? Even if the last update is old, the site might still be worthy of use *if* the content is still valid for your research.

Relevancy to Your Research: Primary versus Secondary Sources

Some research assignments require the use of primary (original) sources. Materials such as raw data, diaries, letters, manuscripts, and original accounts of events can be considered primary material. In most cases, these historical documents are no longer copyrighted. The Web is a great source for this type of resource.

Information that has been analyzed and previously interpreted is considered a secondary source. Sometimes secondary sources are more appropriate than primary sources. If, for example, you are asked to analyze a topic or to find an analysis of a topic, a secondary source of an analysis would be most appropriate. Ask yourself the following questions to determine whether the Web site is relevant to your research:

1. Is it a primary or secondary source?
2. Do you need a primary source?
3. Does the assignment require you to cite different types of sources? For example, are you supposed to use at least one book, one journal article, and one Web page?

You need to think critically, both visually and verbally, when evaluating Web sites. Because Web sites are designed as multimedia hypertexts, nonlinear texts, visual elements, and navigational tools are added to the evaluation process.

Help in Evaluating Web Sites. One shortcut to finding high-quality Web sites is using subject directories and meta-sites, which select the Web sites they index by similar evaluation criteria to those just described. If you want to learn more about evaluating Web sites, many colleges and universities provide sites that help you evaluate Web resources. The following list contains some excellent examples of these evaluation sites:

- Evaluating Quality on the Net—Hope Tillman, Babson College
  www.hopetillman.com/findqual.html
- Critical Web Evaluation—Kurt W. Wagner, William Paterson University of New Jersey
  euphrates.wpunj.edu/faculty/wagnerk/
- Evalation Criteria—Susan Beck, New Mexico State University
  lib.nmsu.edu/instruction/evalcrit.html
- A Student's Guide to Research with the WWW
  www.slu.edu/departments/english/research/
- Evaluating Web Pages: Questions to Ask & Strategies for Getting the Answers
  www.lib.berkeley.edu/TeachingLib/Guides/Internet/EvalQuestions.html

Critical Evaluation Web Sites

WEB SITE AND URL	SOURCE
Critical Thinking in an Online World **www.library.ucsb.edu/untangle/jones.html**	*Paper from "Untangling the Web" 1996*
Educom Review: Information Liberal **www.educause.edu/pub/er/review/reviewArticles/31231.html**	*EDUCAUSE Literacy as a Liberal Art (1996 article)*
Evaluating Information Found on the Internet **MiltonsWeb.mse.jhu.edu/research/education/net.html**	*University of Utah Library*
Evaluating Web Sites **www.lib.purdue.edu/InternetEval**	*Purdue University Library*
Evaluating Web Sites **www.lehigh.edu/~inref/guides/evaluating.web.html**	*Lehigh University*
ICONnect: Curriculum Connection's Overview **www.ala.org/ICONN/evaluate.html**	*American Library Association's technology education initiative*
Kathy Schrock's ABC's of Web Site Evaluation **www.kathyschrock.net/abceval/**	*Author's Web site*
Kids Pick the best of the Web "Top 10: Announced" **www.ala.org/news/topkidpicks.html**	*American Library Association initiative underwritten by Microsoft (1998)*
Resource Selection and Information Evaluation **alexia.lis.uiuc.edu/~janicke/InfoAge.html**	*Univ of Illinois, Champaign-Urbana (Librarian)*
Testing the Surf: Criteria for Evaluating Internet Information Sources **info.lib.uh.edu/pr/v8/n3/smit8n3.html**	*University of Houston Libraries*
Evaluating Web Resources **www2.widener.edu/Wolfgram-Memorial-Library/webevaluation/webeval.htm**	*Widener University Library*

(continued)

WEB SITE AND URL	SOURCE
UCLA College Library Instruction: Thinking Critically about World Wide Web Resources **www.library.ucla.edu/libraries/ college/help/critical/**	*UCLA Library*
UG OOL: Judging Quality on the Internet **www.open.uoguelph.ca/resources/ skills/judging.html**	*University of Guelph*
Web Evaluation Criteria **lib.nmsu.edu/instruction/ evalcrit.html**	*New Mexico State University Library*
Web Page Credibility Checklist **www.park.pvt.k12.md.us/academics/ research/credcheck.htm**	*Park School of Baltimore*
Evaluating Web Sites for Educational Uses: Bibliography and Checklist **www.unc.edu/cit/guides/irg-49.html**	*University of North Carolina*
Evaluating Web Sites **www.lesley.edu/library/guides/ research/evaluating_web.html**	*Lesley University*

Tip: Can't seem to get a URL to work? If the URL doesn't begin with www, you may need to put the http:// in front of the URL. Usually, browsers can handle URLs that begin with www without the need to type in the "http://" but if you find you're having trouble, add the http://.

Documentation Guidelines for Online Sources

Your Citation for Exemplary Research

There's another detail left for us to handle—the formal citing of electronic sources in academic papers. The very factor that makes research on the Internet exciting is the same factor that makes referencing these sources challenging: their dynamic nature. A journal article exists, either in print or on microfilm, virtually forever. A document on the Internet can come, go, and change without warning. Because the purpose of citing sources is to allow another scholar to retrace your argument, a good citation allows a reader to obtain information from your primary sources, to the extent possible. This means you need to include not only information on when a source was posted on the Internet (if available) but also when you obtained the information.

The two arbiters of form for academic and scholarly writing are the Modern Language Association (MLA) and the American Psychological Association (APA); both organizations have established styles for citing electronic publications.

MLA Style

In the fifth edition of the *MLA Handbook for Writers of Research Papers,* the MLA recommends the following formats:

- **URLs:** URLs are enclosed in angle brackets (<>) and contain the access mode identifier, the formal name for such indicators as "http" or "ftp." If a URL must be split across two lines, break it only after a slash (/). Never introduce a hyphen at the end of the first line. The URL should include all the parts necessary to identify uniquely the file/document being cited.

    ```
    <http://www.csun.edu/~rtvfdept/home/
    index.html>
    ```

- **An online scholarly project or reference database:** A complete "online reference contains the title of the project or database (underlined); the name of the editor of the project or database (if given); electronic publication information, including version number (if relevant and if not part of the title), date of electronic publication or latest update, and name of any sponsoring institution or organization; date of access; and electronic address.

    ```
    The Perseus Project. Ed. Gregory R. Crane.
        Mar. 1997. Department of Classics,
        Tufts University. 15 June 1998
        <http://www.perseus.tufts.edu/>.
    ```

If you cannot find some of the information, then include the information that is available. The MLA also recommends that you print or download electronic documents, freezing them in time for future reference.

- **A document within a scholarly project or reference database:** It is much more common to use only a portion of a scholarly project or database. To cite an essay, poem, or other short work, begin this citation with the name of the author and the title of the work (in quotation marks). Then, include all the information used when citing a complete online scholarly project or reference database, however, make sure you use the URL of the specific work and not the address of the general site.

> Cuthberg, Lori. "Moonwalk: Earthlings' Finest Hour." Discovery Channel Online. 1999. Discovery Channel. 25 Nov. 1999 <http://www.discovery.com/indep/newsfeatures/moonwalk/challenge.html>.

- **A professional or personal site:** Include the name of the person creating the site (reversed), followed by a period, the title of the site (underlined), or, if there is no title, a description such as Home page (such a description is neither placed in quotes nor underlined). Then, specify the name of any school, organization, or other institution affiliated with the site and follow it with your date of access and the URL of the page.

> Packer, Andy. Home page. 1Apr. 1998 <http://www.suu.edu/~students/Packer.htm>.

Some electronic references are truly unique to the online domain. These include email, newsgroup postings, MUDs (multiuser domains) or MOOs (multiuser domains, object-oriented), and IRCs (Internet Relay Chats).

Email. In citing email messages, begin with the writer's name (reversed) followed by a period, then the title of the message (if any) in quotations as it appears in the subject line. Next comes a description of the message, typically "Email to," and the recipient (e.g., "the author"), and finally the date of the message.

> Davis, Jeffrey. "Web Writing Resources." Email to Nora Davis. 3 Jan. 2000.
>
> Sommers, Laurice. "Re: College Admissions Practices." Email to the author. 12 Aug. 1998.

List Servers and Newsgroups. In citing these references, begin with the author's name (reversed) followed by a period. Next include the title of the document (in quotes) from the subject line, followed by the words "Online posting" (not in quotes). Follow this with the date of posting. For list servers, include the date of access, the name of the list (if known), and the online address of the list's moderator or administrator. For newsgroups, follow "Online posting" with the date of posting, the date of access, and the name of the newsgroup, prefixed with "news:" and enclosed in angle brackets.

Applebaum, Dale. "Educational Variables." Online posting. 29 Jan. 1998. Higher Education Discussion Group. 30 Jan. 1993 <jlucidoj@unc.edu>.

Gostl, Jack. "Re: Mr. Levitan." Online posting. 13 June 1997. 20 June 1997 <news:alt.edu.bronxscience>.

MUDs, MOOs, and IRCs. Begin with the name of the speaker(s) followed by a period. Follow with the description and date of the event, the forum in which the communication took place, the date of access, and the online address. If you accessed the MOO or MUD through telnet, your citation might appear as follows:

Guest. Personal interview. 13 Aug. 1998. <telnet://du.edu:8888>.

For more information on MLA documentation style for online sources, check out their Web site at http://www.mla.orgstyle/sources.htm.

APA Style

The newly revised *Publication Manual of the American Psychological Association* (5th ed.) now includes guidelines for Internet resources. The manual recommends that, at a minimum, a reference of an Internet source should provide a document title or description, a date (either the date of publication or update or the date of retrieval), and an address (in Internet terms, a uniform resource locator, or URL). Whenever possible, identify the authors of a document as well. It's important to remember that, unlike the MLA, the APA does not include temporary or transient sources (e.g., letters, phone calls, etc.) in its "References" page, preferring to handle them in the text. The general suggested format is as follows:

Online periodical:

Author, A. A., Author, B. B., & Author, C. C. (2000). Title of article. *Title of Periodical, xx,* xxxxx. Retrieved month, day, year, from source.

Online document:

Author, A. A. (2000). *Title of work.* Retrieved month, day, year, from source.

Some more specific examples are as follows:

FTP (File Transfer Protocol) Sites. To cite files available for downloading via FTP, give the author's name (if known), the publication date (if available and if different from the date accessed), the full title of the paper (capitalizing only the first word and proper nouns), the date of access, and the address of the FTP site along with the full path necessary to access the file.

> Deutsch, P. (1991) Archie: An electronic directory service for the Internet. Retrieved January 25, 2000 from File Transfer Protocol: ftp://ftp.sura.net/pub/archie/docs/whatis.archie

WWW Sites (World Wide Web). To cite files available for viewing or downloading via the World Wide Web, give the author's name (if known), the year of publication (if known and if different from the date accessed), the full title of the article, and the title of the complete work (if applicable) in italics. Include any additional information (such as versions, editions, or revisions) in parentheses immediately following the title. Include the date of retrieval and full URL (the http address).

> Burka, L. P. (1993). A hypertext history of multi-user dungeons. *MUDdex*. Retrieved January 13, 1997 from the World Wide Web: http://www.utopia.com/talent/lpb/muddex/essay/

> Tilton, J. (1995). Composing good HTML (Vers. 2.0.6). Retrieved December 1, 1996 from the World Wide Web: http://www.cs.cmu.edu/~tilt/cgh/

Synchronous Communications (MOOs, MUDs, IRC, etc.). Give the name of the speaker(s), the complete date of the conversation being referenced in parentheses, and the title of the session (if applicable). Next, list the title of the site in italics, the protocol and address (if applicable), and any directions necessary to access the work. Last, list the date of access, followed by the retrieval information. Personal interviews do not need to be listed in the References, but do need to be included in parenthetic references in the text (see the APA *Publication Manual*).

Cross, J. (1996, February 27). Netoric's Tuesday
 "cafe: Why use MUDs in the writing classroom?
 MediaMoo. Retrieved March 1, 1996 from File
 Transfer Protocol: ftp://daedalus.com/pub/
 ACW/NETORIC/catalog

Gopher Sites. List the author's name (if applicable), the year of publication, the title of the file or paper, and the title of the complete work (if applicable). Include any print publication information (if available) followed by the protocol (i.e., gopher://). List the date that the file was accessed and the path necessary to access the file.

Massachusetts Higher Education Coordinating
 Council. (1994). Using coordination and
 collaboration to address change. Retrieved
 July 16, 1999 from the World Wide Web:
 gopher://gopher.mass.edu:170/
 00gopher_root%3A%5B_hecc%5D_plan

Email, Listservs, and Newsgroups. Do not include personal email in the list of References. Although unretrievable communication such as email is not included in APA References, somewhat more public or accessible Internet postings from newsgroups or listservs may be included. See the APA *Publication Manual* for information on in-text citations.

Heilke, J. (1996, May 3). Webfolios. Alliance
 for Computers and Writing Discussion List.
 Retrieved December 31, 1996 from the World
 Wide Web: http://www.ttu.edu/lists/acw-l/
 9605/0040.html

Other authors and educators have proposed similar extensions to the APA style. You can find links to these pages at:

www.psychwww.com/resource/apacrib.htm

Remember, "frequently-referenced" does not equate to "correct" "or even "desirable." Check with your professor to see if your course or school has a preference for an extended APA style.

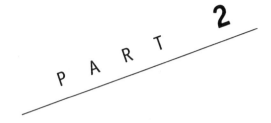

PART 2

ContentSelect

What Is ContentSelect?

http://www.ablongman.com/contentselect

Allyn & Bacon and EBSCO Publishing, leaders in the development of electronic journal databases have exclusively collaborated to develop the ContentSelect Research Database, an online collection of leading scholarly and peer-reviewed journals in the discipline. Students can have unlimited access to a customized, searchable collection of discipline-specific articles from top-tier academic publications.

In addition, new features are especially designed to help you with the research process:

- **Start Writing!** With detailed information on the process of writing a research paper, from finding a topic, to gathering data, using the library, using online sources, and more.
- **Internet Research** and **Resource Links** aggregates links to many of the best sites on the Web, providing more tips and best practices to help you use the Web for research.

- **Citing Sources.** Featuring excerpts from the best-selling book on research paper, this section helps you understand how and when to cite sources, and includes examples of various citation styles.

How to Use ContentSelect

To begin exploring the great resources available in the ContentSelect Research Database Web site:

Step 1: Go to: **http://www.ablongman.com/contentselect**

Step 2: The resources on the home page will help you start the research and writing process and cite your sources. For invaluable research help,

- Click **Citing Sources** to see how to cite materials with these citation styles: MLA, APA, CMS, and CBE.
- Click **Start Writing!** for step-by-step instructions to help you with the process of writing a research paper.
- Click **Resource Links** and **Internet Research** to link to many of the best sites on the Web with tips to help you efficiently use the Web for research.

Step 3: Register! To start using the ContentSelect Research Database, you will need to register using the access code and instructions located on the inside front cover of this guide. You only need to register once—after you register, you can return to ContentSelect at any time, and log in using your personal login name and password.

Step 4: Log in! Type in your login name and password in the spaces provided to access ContentSelect. Then click through the pages to enter the research database, and see the list of disciplines. You can search for articles within a single discipline, or select as many disciplines as you want! To see the list of journals included in any database, just click the "**complete title list**" link located next to each discipline—check back often, as this list will grow throughout the year!

Step 5: To begin your search, simply select your discipline(s), and **click "Enter"** to begin your search. For tips and detailed search instructions, please visit the "ContentSelect Search Tips" section included in this guide.

For more help, and search tips, click the Online Help button on the right side of your screen.

Go to **www.ablongman.com/contentselect** now, to discover the easiest way to start a research paper!

ContentSelect Search Tips

Searching for articles in ContentSelect is easy! Here are some tips to help you find articles for your research paper.

Tip 1: **Select a discipline.** When you first enter the Content-Select Research Database, you will see a list of disciplines. To search within a single discipline, click the name of the discipline. To search in more than one discipline, click the box next to each discipline and click the **ENTER** button.

Basic Search

The following tips will help you with a Basic Search.

Tip 2: **Basic Search.** After you select your discipline(s), you will go to the Basic Search Window. Basic Search lets you search for articles using a variety of methods. You can select from: Standard Search, Match All Words, Match Any Words, or Match Exact Phrase. For more information on these options, click the Search Tips link at any time!

Tip 3: **Using AND, OR, and NOT** to help you search. In Standard Search, you can use AND, OR and NOT to create a very broad or very narrow search:

- **AND** searches for articles containing all of the words. For example, typing **education AND technology** will search for articles that contain **both** education AND technology.

- **OR** searches for articles that contains at least one of the terms. For example, searching for **education OR technology** will find articles that contain either education OR technology.

- **NOT** excludes words so that the articles will not include the word that follows "NOT." For example, searching for **education NOT technology** will find articles that contain the term education but NOT the term technology.

Tip 4: **Using Match All Words.** When you select the Match All Words option, you do not need to use the word AND—you will automatically search for articles that only contain all of the words. The order of the search words entered in does not matter. For example, typing **education technology** will search for articles that contain **both** education AND technology.

Tip 5: **Using Match Any Words.** After selecting the "Match Any Words" option, type words, a phrase, or a sentence in the window. ContentSelect will search for articles that contain any of the terms you typed (but will not search for words such as **in** and **the**). For example, type **rising medical costs in the United States** to find articles that contain *rising, medical, costs, United,* or *States*. To limit your search to find articles that contain exact terms, use *quotation marks*—for example, typing "United States" will only search for articles containing "United States."

Tip 6: **Using Match Exact Phrase.** Select this option to find articles containing an exact phrase. ContentSelect will search for articles that include all the words you entered, exactly as you entered them. For example, type **rising medical costs in the United States** to find articles that contain the exact phrase "rising medical costs in the United States."

Guided Search

The following tips will help you with a Guided Search.

Tip 7: To switch to a Guided Search, click the **Guided Search** tab on the navigation bar, just under the EBSCO Host logo. The *Guided Search Window* helps you focus your search using multiple text boxes, Boolean operators (AND, OR, and NOT), and various search options.

To create a search:

- Type the words you want to search for in the Find field.
- Select a field from the drop-down list. For example: AU-Author will search for an author. For more information on fields, click Search Tips.
- Enter additional search terms in the text boxes (optional), and select *and, or, not* to connect multiple search terms (see Tip 3 for information on *and, or,* and *not*).
- Click **Search.**

Expert Search

The following tips will help you with an Expert Search.

Tip 8: To switch to an Expert Search, click the **Expert Search** tab on the navigation bar, just under the EBSCO Host logo. The *Expert Search Window* uses your keywords and search history search for articles. Please note, searches run from the Basic or Guided Search Windows are not saved to the History File used by the Expert Search Window—only Expert Searches are saved in the history.

Tip 9: Expert Searches use **Limiters** and **Field Codes** to help you search for articles. For more information on Limiters and Field Codes, click Search Tips.

Explore all the search options available in ContentSelect! For more information and tips, click the Online Help button, located on the right side of every page.

Glossary

Your Own Private Glossary

The Glossary in this book contains reference terms you'll find useful as you get started on the Internet. After a while, however, you'll find yourself running across abbreviations, acronyms, and buzzwords whose definitions will make more sense to you once you're no longer a novice (or "newbie"). That's the time to build a glossary of your own. For now, the Webopedia gives you a place to start.

alias A simple email address that can be used in place of a more complex one.

AVI Audio Video Interleave. A video compression standard developed for use with Microsoft Windows. Video clips on the World Wide Web are usually available in both AVI and QuickTime formats.

bandwidth Internet parlance for capacity to carry or transfer information such as email and Web pages.

browser The computer program that lets you view the contents of Web sites.

client A program that runs on your personal computer and supplies you with Internet services, such as getting your mail.

cyberspace The whole universe of information that is available from computer networks. The term was coined by science fiction writer William Gibson in his novel *Neuromancer*, published in 1984.

DNS See *domain name server*.

domain A group of computers administered as a single unit, typically belonging to a single organization such as a university or corporation.

domain name A name that identifies one or more computers belonging to a single domain. For example, "apple.com".

domain name server A computer that converts domain names into the numeric addresses used on the Internet.

download Copying a file from another computer to your computer over the Internet.

email Electronic mail.

emoticon A guide to the writer's feelings, represented by typed characters, such as the Smiley :-). Helps readers understand the emotions underlying a written message.

FAQs Frequently Asked Questions

flame A rude or derogatory message directed as a personal attack against an individual or group.

flame war An exchange of flames (see above).

ftp File Transfer Protocol, a method of moving files from one computer to another over the Internet.

home page A page on the World Wide Web that acts as a starting point for information about a person or organization.

hypertext Text that contains embedded *links* to other pages of text. Hypertext enables the reader to navigate between pages of related information by following links in the text.

LAN Local Area Network. A computer network that is located in a concentrated area, such as offices within a building.

link A reference to a location on the Web that is embedded in the text of the Web page. Links are usually highlighted with a different color or underlined to make them easily visible.

listserv Strictly speaking, a computer program that administers electronic mailing lists, but also used to denote such lists or discussion groups, as in "the writer's listserv."

lurker A passive reader of an Internet *newsgroup* or *listserv*. A lurker reads messages, but does not participate in the discussion by posting or responding to messages.

mailing list A subject-specific automated email system. Users subscribe and receive email from other users about the subject of the list.

modem A device for connecting two computers over a telephone line.

newbie A new user of the Internet.

newsgroup A discussion forum in which all participants can read all messages and public replies between the participants.

plug-in A third-party software program that will lend a Web browser (Netscape, Internet Explorer, etc.) additional features.

quoted Text in an email message or newsgroup posting that has been set off by the use of vertical bars or > characters in the left-hand margin.

search engine A computer program that will locate Web sites or files based on specified criteria.

secure A Web page whose contents are encrypted when sending or receiving information.

server A computer program that moves information on request, such as a Web server that sends pages to your browser.

Smiley See *emoticon*.

snail mail Mail sent the old fashioned way: Write a letter, put it in an envelope, stick on a stamp, and drop it in the mailbox.

spam Spam is to the Internet as unsolicited junk mail is to the postal system.

URL Uniform Resource Locator: The notation for specifying addresses on the World Wide Web (e.g. http://www.abacon.com or ftp://ftp.abacon.com).

Usenet The section of the Internet devoted to *newsgroups*.

Web browser A program used to navigate and access information on the World Wide Web. Web browsers convert html coding into a display of pictures, sound, and words.

Web page All the text, graphics, pictures, and so forth, denoted by a single URL beginning with the identifier "http://".

Web site A collection of World Wide Web pages, usually consisting of a home page and several other linked pages.